The Unique Force Pattern

STEMIONICS
GRAND UNIFIED THEORY

STEMIONICS

The Unique Force Pattern
___

> I present you the story.
> You can imagine its details.
> But only God understands its meaning.

STEMIONICS

1. THE STORY OF STEMIONICS

Imagine a garniture of subway that rushes through the underground gallery in which the only lighted places are the stations.

During its travel no one can get on or get off, but on board of the subway, passengers interact continuously: they exchange opinions, they like or dislike each other, they nap or they fuss intensely.

At the station the doors open, there is a lot of commotion, some passengers get on, other get off, some take the place of others, then the doors close again and the travel continues to the next station.

There are different people in the subway now, but they behave alike, travelling together through the dark gallery toward the next station.

Each one of the passengers seems to be driven by a certain impulse, but - on the whole - each one, just like the garniture of the subway, moves in its wholeness, for only one reason: the vital force.

Because of it, each individual lives his life as part of the human civilization, interacting continuously and - directly or indirectly - everyone influences everyone.

This is the story of a subway.

Now imagine this:

The dark subway gallery is the dark and cold absolute potential void;

The subway is the Universe, which transports matter and energy with its Gravitational Field;

The passengers are the Energy Quanta, which interact through Electromagnetic Forces, combining themselves constantly between stations;

The stations represent singularities (Nullons) in which the energy transported through a universe suffers a state transformation and then leaves to a different Universe.

If you are in a Universe, you are a prisoner of the gravitation that transports you along with the energy of the entire Universe.

If you are in the Nullon, the gravity does not affect you, but you can see the total balance - the quantity of energy that comes in and out of a Universe.

So, although the causes of the interactions seem equally different, just like the reasons of the passengers from the subway, they actually are particular cases of a Unique Universal Force.

Because of it, each Quantum interacts continuously with any other Quantum, influencing each other in a continuous recuperative pattern, duplicated at every level of the energy existence: The Standard Stemionic Pattern of the Unified Force Field.

This is the story of Stemionics.

Why "STEMIONICS"?

The pattern is based on phenomena having origin in a geometrical place called STEMION.

STEMION means acronym of:

Space, Time, Energy, Mass, Information Originar Node.

2. THE OBJECTIVES OF THE PATTERN

Stemionics proposes a complex model of Grand Unified Theory disambiguation which wants to show that:

There is a Unique Universal Force that manifests itself as a continuous interraction between all the energy quanta.

The Unique Universal Force is the Primordial Cause for the existence of an unlimited number of Universes located on a Field of Potential Universes.

All the phenomena from any Universe are a consequence of this force.

These Universes "flow" constantly one into the other, each Universe being an energy transporter for the next Universe.

Each Universe represents a potential dump located on a void amplitude core of a linear circular tensor on which is defined this transport function, and the energy transfer between the universes takes place on the loops of this tensor.

The total balances of the energy and of the impulse of a Universe are void and the total balance of the energies and impulses transfers between all the potential Universes is void.

All the subsystems of a Universe evolve in the sense of balancing the total potential, but also the relative one.

Stemionics asserts that no matter which level the observer is situated in the interior of a Universe,, he participates in the physical unsteady phenomena, being transported with them, "captive" by the gravity (the transport energy) toward the next loop.

Meanwhile, the observer from the loop is not affected by the physical unsteady phenomena from the interior of the Universe or by its gravitational field. He sees the total balance: its energy "quantum" of transfer, which represents a state electromagnetic energy.

The Unique Universal Force defines, a unique pattern, repeatable at every scale, for the Micro Universe, Macro Universe or Mega Universe.

The stemionic pattern can coexist mainly with other physical patterns, having its own relativistic approach and a quantum interpretation which offer a disambiguation of certain approaches of the classical quantum mechanics.

Unlike the classical physical patterns, the stemionic pattern is a pattern whose approach surpasses the spatial-temporary limits of the Universe we live in, which is considered only a particular case from the Field of Potential Universes.

The following will be presented in synthesis.

3. THE STANDARD STEMIONIC PATTERN OF GRAND UNIFIED THEORY

The energy EON elementary particle has a multiple continuous energetic potential in energetic fields;

This energy has te following form:

$$e_{TOTAL} = (...- ie- ...-2e - e + e + 2e+ ... + i e...)$$

It is born from an **universal inergetic field (ORIGINAR DARK ENERGY FIELD)**

formed out of identical origins called STEMION,

having a multiple inergetic potential in the form of:

$$\ddot{i}_{i\ TOTAL} = ...(- i_n\ \ddot{i}_i) + ... + (- i_2\ \ddot{i}_i) + (- i_1\ \ddot{i}_i) + (+ i_1\ \ddot{i}_i) + (+ i_2\ \ddot{i}_i) + ... + (+ i_n\ \ddot{i}_i)...$$

in which the Unique Universal Force is in the form of:

$$\ddot{O}\ddot{O}T\check{U} = \ddot{A}(\frac{1}{2}\sum \ddot{i}\Delta \ddot{i}\ \Theta\ \frac{1}{2}\sum \ddot{i}\Delta \ddot{i}) + 4\ddot{A}(\frac{1}{2}\sum \ddot{i}\Delta \ddot{i}\ \Theta\ \sum \ddot{i}\Delta \ddot{i}).$$

It suffers a continuous transformation from **inergy to energy**

in **pre-energetic prenergetic field (DARK TRANZITORY ENERGY FIELD)** *being a transitory particle – LERON:*

$$\boxed{E = -IF^2}$$

In this prenergetic field the Unique Universal Force is in the form of:

$$\bar{\bar{O}}\bar{\bar{O}}T\check{U} = \bar{A}\left(\frac{1}{2}\sum i\Delta i \ominus \frac{1}{2}\sum i\Delta i\right) + 4\bar{A}\left(\frac{1}{2}\sum i\Delta i \ominus \sum i\Delta i\right).$$

It evolves on the **universal energy field (BARYONIC ENERGY FIELD)**

in which the Unique Universal Force between the elementary energy particles EON is in the form of:

$$\dot{E}\dot{E}T = \frac{\dot{A}}{D^2}\left(\frac{1}{2}\sum e\Delta e \ominus \frac{1}{2}\sum e\Delta e\right) + 4\frac{\dot{A}}{D^2}\left(\frac{1}{2}\sum e\Delta e \ominus \frac{1}{2}\sum e\Delta e\right).$$

Where term:

$$\frac{\dot{A}}{D^2}\left(\frac{1}{2}\sum e\Delta e \ominus \frac{1}{2}\sum e\Delta e\right)$$

represent the interraction of **BARYONIC MATTER**

and the term:

$$4\frac{\bar{A}}{D^2}\left(\frac{1}{2}\sum e\Delta e \;\Theta\; \frac{1}{2}\sum e\Delta e\right)$$

represent the interraction of **DARK MATTER.**

*It suffers a transformation from **energy to inergy***

*on an **post-energetic prenergetic field** (**DARK TRANZITORY ENERGY FIELD**),*

being a transitory particle – LERON,

in which the Unique Universal Force between the elementary energy particles EON is in the form of:

$$-\bar{\bar{O}}\bar{\bar{O}}T\check{U} = -\bar{A}\left(\frac{1}{2}\sum i\Delta i \;\Theta\; \frac{1}{2}\sum i\Delta i\right) - 4\bar{A}\left(\frac{1}{2}\sum i\Delta i \;\Theta\; \sum i\Delta i\right).$$

*And it ends its universal evolution in the **universal inergetic field** (**ORIGINAR DARK ENERGY FIELD**)*

The Unique Force Pattern

in which the Unique Universal Force is the same as in the originar one, in the form of:

$$\ddot{\ddot{O}}T\breve{U} = \ddot{A}\left(\frac{1}{2}\sum \ddot{i}\Delta \ddot{i} \ \Theta \ \frac{1}{2}\sum \ddot{i}\Delta \ddot{i}\right) + 4\ddot{A}\left(\frac{1}{2}\sum \ddot{i}\Delta \ddot{i} \ \Theta \ \sum \ddot{i}\Delta \ddot{i}\right).$$

All the four fields exert on the EON particle continuous interactions,

which lead us to the conclusion that the Total Unique Universal Force

exerted on the energetic EON particles in the Universal Energy Field

and on the pre-energetic and post-energetic field in their vicinity is in the form of:

$$\dot{\dot{E}}E_{TOTAL} = \ddot{\ddot{O}}T\breve{U} + \bar{\bar{O}}\bar{O}T\breve{U} - \bar{\bar{O}}\bar{O}T\breve{U} + \dot{E}\dot{E}T,$$

Where:

$\ddot{\ddot{O}}T\breve{U}$ – Total Unique Universal Force of the inergetic field,

13

$\pm \bar{O}\bar{O}T\check{U}$ - *Total Unique Universal Force of the prenergetic fields,*

ĖĖT - Local Unique Universal Force of the energetic field.

We observe the distribution of matter interractions being most probably:

~ 75% - DARK ENERGY – (Λ_{CMD}) - responsible for accelerated expansion of the Universe,

~ 20% - DARK MASS – (WIMP) - responsible for creating a halo gravitational around a massive body equivalent to its diameter,

~ 5% - BARYONIC MASS.

There exist also a small percentage of inergetic or prenergetic particles classical defined as **neutrino**.

For two identically energy charged EONs with energetic potential:

$e_1 = e_2$

The Unique Force Pattern

we write the total variation of the force field

by the vector of the relative direction of movement of the two EONs

as bellow:

$$\frac{\partial}{\partial Di}\dot{E}\dot{E}\text{ TOTAL} =$$

$$-\frac{\partial}{\partial Di}\frac{1}{\ddot{u}^2}(\frac{1}{2}\sum i\Delta i \ominus \frac{1}{2}\sum i\Delta i)$$

$$-4\frac{\partial}{\partial Di}\frac{1}{\ddot{u}^2}(\frac{1}{2}\sum i\Delta i \ominus \sum i\Delta i)$$

$$+\frac{\partial}{\partial Di}\frac{1}{\ddot{u}^2}(\frac{1}{2}\sum \bar{i}\Delta \bar{i} \ominus \frac{1}{2}\sum \bar{i}\Delta \bar{i})$$

$$+4\frac{\partial}{\partial Di}\frac{1}{\ddot{u}^2}(\frac{1}{2}\sum \bar{i}\Delta \bar{i} \ominus \sum \bar{i}\Delta \bar{i})$$

$$+\frac{\partial}{\partial Di}\frac{1}{\ddot{u}^2}(\frac{1}{2}\sum e\Delta e \ominus \frac{1}{2}\sum e\Delta e)$$

$$+4\frac{\partial}{\partial Di}\frac{1}{\ddot{u}^2}(\frac{1}{2}\sum e\Delta e \ominus \frac{1}{2}\sum e\Delta e)$$

$$-\frac{\partial}{\partial Di}\frac{1}{\ddot{u}^2}(\frac{1}{2}\sum \bar{i}\Delta \bar{i} \ominus \frac{1}{2}\sum \bar{i}\Delta \bar{i})$$

$$-4\frac{\partial}{\partial Di}\frac{1}{\ddot{u}^2}(\frac{1}{2}\sum \bar{i}\Delta \bar{i} \ominus \sum \bar{i}\Delta \bar{i}).$$

For any finite sector we can observe only local variable interactions

we write equation of the local variation of the force field as bellow:

$$\frac{\partial}{\partial Di}\dot{E}\dot{E} =$$

$$-\frac{\partial}{\partial Di}\ddot{A}\left(\frac{1}{2}\sum i\Delta i \ \ \Theta \frac{1}{2}\sum i\Delta i\right)$$

$$-4\frac{\partial}{\partial Di}\ddot{A}\left(\frac{1}{2}\sum i\Delta i \ \ \Theta \frac{1}{2}\sum i\Delta i\right)$$

$$+\frac{\partial}{\partial Di}\bar{A}\left(\frac{1}{2}\sum \bar{i}\Delta \bar{i} \ \Theta \frac{1}{2}\sum \bar{i}\Delta \bar{i}\right)$$

$$+4\frac{\partial}{\partial Di}\bar{A}\left(\frac{1}{2}\sum \bar{i}\Delta \bar{i} \ \Theta \frac{1}{2}\sum \bar{i}\Delta \bar{i}\right)$$

$$+\frac{\partial}{\partial Di}\dot{A}\left(\frac{e1}{D} \Theta \frac{e2}{D}\right)$$

$$+4\frac{\partial}{\partial Di}\dot{A}\left(\frac{e1}{D} \Theta \frac{e2}{D}\right)$$

$$-\frac{\partial}{\partial Di}\bar{A}\left(\frac{1}{2}\sum \bar{i}\Delta \bar{i} \ \Theta \frac{1}{2}\sum \bar{i}\Delta \bar{i}\right)$$

$$-4\frac{\partial}{\partial Di}\bar{A}\left(\frac{1}{2}\sum \bar{i}\Delta \bar{i} \ \Theta \ \frac{1}{2}\sum \bar{i}\Delta \bar{i}\right).$$

Therefore, we have:

a variation of the Unique Universal Force between the two EON

in the form of the gradient vector $\vec{\nabla}\dot{E}\dot{E}_{TOTAL}$

as bellow:

The Unique Force Pattern

> ***The Unique Equation in a relative form:***
> ***In the Universal Energetic Field the total relative variation of the Unique Universal Force between two EONs is in the form:***
> $$\nabla \dot{E}\dot{E}_{TOTAL} =$$
> $$\nabla \ddot{O}\ddot{O} T\check{U} + \nabla \bar{O}\bar{O}T\check{U} + \nabla \dot{E}\dot{E}T\check{U} - \nabla \bar{O}\bar{O}T\check{U}$$

We can write the equation dependent on

the scalar product between the gradient $\nabla \dot{E}\dot{E}_{TOTAL}$

and the vector of the tension curve on which evolves the particle in the energy field.

Therefore, we will obtain the directional derivative of $\dot{E}\dot{E}_{TOTAL}$

after the tension curve on which evolves the particle in the energy field.

This is the most exact form of the absolute variation of the Unique Universal Force

in proportion to the origin and the spatial coordinates of the universal energy field.

> **The Unique Equation in an absolute form:**
> **In the Universal Energy Field the total absolute variation of the Unique Universal Force is in the form:**
>
> $\langle \vec{\nabla}\dot{E}\dot{E}_{TOTAL}, \underline{R}_i \rangle = \langle \ddot{O}\ddot{O}T\check{U}, \underline{R}_i \rangle$
> $+ \langle \vec{\nabla}\bar{O}\bar{O}T\check{U}, \underline{R}_i \rangle$
> $+ \langle \vec{\nabla}\dot{E}\dot{E}T\check{U}, \underline{R}_i \rangle$
> $+ \langle -\vec{\nabla}\bar{O}\bar{O}T\check{U}, \underline{R}_i \rangle$

We write the two components of the Unique Universal Force in their quantum correspondent form:

$$\bar{O}\bar{O}T = \frac{H v_{\tilde{F}}}{d}.$$

For a local sector from the energy field (for example the solar system),

in which $\dot{A}$ is relatively constant:

$$\frac{\partial}{\partial D_i} \dot{E}\dot{E}_{TOTAL} = -\frac{H v_{\tilde{F}}}{d} + \frac{H v_{\tilde{F}}}{d} + \frac{4\dot{A}e^2}{D_i^3}$$

in the limited form:

$$\frac{\partial}{\partial D_i} \dot{E}\,\dot{E}_{TOTAL} = \frac{4\dot{A}e^2}{D_i^3}.$$

The Unique Force Pattern

So:

$$\vec{V\dot{E}\dot{E}}_{TOTAL} = (\frac{4\dot{A}e^2}{D_{i1}^3} \dots , \frac{4\dot{A}e^2}{D_{in}^3}).$$

Therefore:

$$\vec{V\dot{E}\dot{E}}_{TOTAL} = \vec{V\dot{E}\dot{E}}.$$

We can see that

the balances of the EON's interaction with the pre-energetic exterior fields cancel each other.

So they do not influence the total balance of the Unique Universal Force

exerted on the EON particle in the interior of the energy field.

These components of the Unique Universal Force

are responsible only for the "loading", and "unloading" of the energy charge, respectively.

We can state:

> **The Unique Equation simplified in a relative form:**
>
> In the Universal Energy Field, the total variation of the Unique Universal Force exerted between two EONs is equal to the variation of the direct interaction between the EON particles on the direction of the relative movement between the two EONs.
>
> $$\vec{VEE}_{TOTAL} = \vec{VEE}$$

And:

> **The Unique Equation simplified in a absolute form:**
>
> In the Universal Energy Field the absolute variation of the direct interaction between the EON particles on the direction of the tension curve on which evolve the two EONs is in the form:
>
> $$\langle \vec{VEE}_{TOTAL}, \underline{R}_i \rangle = \langle \vec{VEE}, \underline{R}_i \rangle$$

In a local subsystem, of the universal energy field, in which $\dot{A}$ is constant:

$$\vec{VE}_{TOTAL} D_i = \vec{VE} D_i.$$

Therefore this is dependent only on the potential energy variation/of transport

of the energy particle in the energy field:

$$\frac{\partial}{\partial D_i} \dot{E}_{TOTAL} D_i = \frac{\partial}{\partial D_i} \dot{E} D_i.$$

So:

$\nabla E_{TOTAL} = \nabla E_t$ *(XII.1.24.).*

We can write the simplified unique equation in relation to the variation of the energy of an EON.

> **The Unique Equation in the energetic form:**
> **The variation of the energy of a EON energy particle in an energy field is proportional to the variation of the transport energy of the EON particle on that energy field.**
> $\nabla E_{TOTAL} = \nabla E_t$

The Unique Equation admits a simple, invariable form:

> **The Unique Equation in the invariable form:** In the Universal Energy Field, the total Unique Universal Force exerted between two EONs is equal to equation: $\dot{E}\dot{E} = A \dfrac{e^2}{D^2}$.

Decomposing $\underline{R}_i$ as carrying space stemionic, we have:

$$\underline{R}_i = \underline{T}_i \,|\psi(t)\rangle \,,$$

$\underline{T}_i = \theta \underline{T}_i$ - spatial displacement vector

and:

$|\psi(t)\rangle = \underline{\varphi r} T_i$ - wave vector (vector KET), for pulsation/circular polarized radiation (normalized to 1).

We rewrite The Unique Equation in the absolute form:

$$E_t |\psi(t)\rangle \cdot \underline{T}_i = \hat{H} |\psi(t)\rangle \cdot \underline{T}_i.$$

For a number j of paricles we define a tensor of space travel:

$$\underline{T}_{ij} = \sum_{i=0}^{j} \underline{T}_i.$$

We formulate tensorial relationship:

$$\dot{A}E_t \underline{T}_{ij} = G_{ij} + \Lambda g_{ij} = kT_{ij}$$

*corresponding to a **constant toroidal shape of the Universe having a Central Nullon as in Annex 1**.*

We write the set of equations :

$$\begin{cases} E_t|\psi(t)\rangle = (\hat{H} + Hv\mathcal{F} - Hv\mathcal{F})|\psi(t)\rangle \\ \dot{A}E_t \underline{T}_{ij} = kT_{ij} \end{cases}$$

We interpret the set of equations defining:

The Unique Equation in the absolute form in quantum-relativistic equivalence - Grand Unified Equation:
$$\dot{A}E^2 \underline{R}_i \equiv k \cdot \hat{H} \cdot |\psi(t)\rangle \otimes T_{ij}$$

Writing ket vector in Stokes vector and writing stress tensor

in minkowskian the four-dimensional space:

for right circularly polarized electromagnetic wave/light

we have:

Grand Unified Equation in matrix form:

$$\dot{A}E^2\underline{R}_i \equiv k \cdot \hat{H} \begin{pmatrix} 1 \\ 0 \\ 0 \\ 1 \end{pmatrix} \otimes \begin{pmatrix} p & 0 & 0 & 0 \\ 0 & p & 0 & 0 \\ 0 & 0 & p & 0 \\ 0 & 0 & 0 & p \end{pmatrix}.$$

---

and for left circularly polarized electromagnetic wave/light

we have:

Grand Unified Equation in matrix form:

$$\dot{A}E^2\underline{R}_i \equiv k \cdot \hat{H} \begin{pmatrix} 1 \\ 0 \\ 0 \\ -1 \end{pmatrix} \otimes \begin{pmatrix} p & 0 & 0 & 0 \\ 0 & p & 0 & 0 \\ 0 & 0 & p & 0 \\ 0 & 0 & 0 & p \end{pmatrix}.$$

Equation has at least a few solutions as the bellow ones:

1. For the small section of universal geodesics:

$t = t'$

and the tensor T_{ab} is normalized as unitary tensor:

$$T_{ab} = \begin{pmatrix} 1 & 0 & 0 & 0 \\ 0 & 1 & 0 & 0 \\ 0 & 0 & 1 & 0 \\ 0 & 0 & 0 & 1 \end{pmatrix}.$$

The equation verify the local variation of the quantum energy:

$$E\,\Psi(t) = \hat{H}\,\Psi(t).$$

2. For the long section of universal geodesics:

The probability $\psi(t)$ is sensitive close to 1 and $\hat{H}$ is an eigenvalue of the each geodesic.

$$\psi(t) \rightarrow 1,\ \hat{H} = ct.$$

Tensor T_{ab} is defined as:

$$T_{ab} = \begin{pmatrix} p & 0 & 0 & 0 \\ 0 & p & 0 & 0 \\ 0 & 0 & p & 0 \\ 0 & 0 & 0 & p \end{pmatrix}$$

The valid equation is:

$$G_{ij} + \Lambda g_{ij} = kT_{ij}.$$

3. For the observer situated outside of the Universe:

All the events in the Universe are simultaneous:

T = 0 – Linear Time.

We have the ballance values

$$\hat{H} = 0,\ \psi(t) = 1.$$

The tensor T_{ab} is normaliset as null tensor:

$$T_{ab} = \begin{pmatrix} 0 & 0 & 0 & 0 \\ 0 & 0 & 0 & 0 \\ 0 & 0 & 0 & 0 \\ 0 & 0 & 0 & 0 \end{pmatrix}.$$

The valid quantum equation is:

$$E\Psi = \hat{H}\Psi.$$

And the valid relativistic equation is:

$$G_{ij} + \Lambda g_{ij} = kT_{ij}$$

We conclude that:

1. For the small sections of universal geodesics the problem has a quantum nature

2. For the long sections of universal geodesics the problem has a relativistic nature

3. Overall balance of the entire Universe can be defined precisely by both models.

According to Stemionic Model

our Universe is a particular case of Universes

*situated on Matrix A — the **Ring of Potential Universes**.*

which, together with its inverse

*form the **Corp of Potential Universes and Antiuniverses**.*

The Matrix A has the form:

$$
\begin{array}{ccccc}
 & + & - & + & - \\
\ldots\ n^0(n^i e) & n^0(n^i e) & n^0(n^i e) & n^0(n^i e) & \ldots \\
\ldots\ n^1(n^{i-1}e) & n^1(n^{i-1}e) & n^1(n^{i-1}e) & n^1(n^{i-1}e) & \ldots \\
\ldots & \ldots & \ldots & \ldots & \ldots \\
\ldots\ n^{i-2}(n^2 e) & n^{i-2}(n^2 e) & n^{i-2}(n^2 e) & n^{i-2}(n^2 e) & \ldots \\
\ldots\ n^{i-1}(n^1 e) & n^{i-1}(n^1 e) & \boxed{\mathbf{n^{i-1}(n^1 e)}} & n^{i-1}(n^1 e) & \ldots \\
\ldots\ n^i(n^0 e) & n^i(n^0 e) & n^i(n^0 e) & n^i(n^0 e) & \ldots \\
\ldots\ n^{i+1}(n^{-1}e) & n^{i+1}(n^{-1}e) & n^{i+1}(n^{-1}e) & n^{i+1}(n^{-1}e) & \ldots \\
\ldots & \ldots & \ldots & \ldots & \ldots \\
\ldots\ n^{i+i}(n^{-i}e) & n^{i+i}(n^{-i}e) & n^{i+i}(n^{-i}e) & n^{i+i}(n^{-i}e) & \ldots \\
\end{array}
$$

*We have on the column **Quantum Energy function** measurable at any level*

*and on the row the **Energy Transport functions** between universes - tensors ALL$_i$.*

The Unique Force Pattern

For all ALL$_i$ tensors we will have a balance equation for the Corp of Potential Universes and Antiuniverses:

$$ALL = det(A) + det(-A).$$

We can formulate:

> **Law 0 of Stemionics:**
> **Overall balance of the interactions on the Corp of Potential Universes and Antiuniverses is zero.**
> **ALL = 0**

The most used notations:

$\dot{A} = \frac{1}{\dot{u}^2}$ – *the Unique Universal Constant* $(\dot{A} = Gc^4)$,

$A = 5\dot{A}$,

$\ddot{A} = \frac{1}{\ddot{u}^2}$ - *the Unique Inergetic Constant*,

$\bar{A} = \frac{1}{\bar{u}^2}$ - *the Unique Prenergetic Constant*,

d – *the distance of the particle's evolution on the pre-energetic field – an intermediary field between energetic and inergetic field,*

D – *the distance between the EONs in the energy field,*

Δ – *delus* – *the operator of the infinite-dimensional assembly:*

$z \Delta y = \{ -i(z+y)\ldots, -(z+y), \ldots 0, z+y, \ldots i(z+y)\}$,

$\sum \ddot{i} \Delta \ddot{i} = (-n)\ddot{i} + \ldots + (-2)\ddot{i} + (-1)\ddot{i} + 0 + (+1)\ddot{i} + (+2)\ddot{i} + \ldots + (+n)\ddot{i}$,

e – the energy of the EON elementary energy particle,

E - the total energy of a finite energy system

E_t – *the transport energy of a finite energy system,*

EON – Energy Originar Node,

I – the correspondent inergy for energy E of a system,

G – gravitational constant,

G_{ij} – *Einstein tensor,*

g_{ij} – *metric tensor,*

Hv_T – *the energy of the particle in its quantum form,*

$\hat{H}$ – *the hamiltonian,*

ï – inergy – state of the energetic fields,

$k = \frac{8\pi G}{c^4}$ – *universal constant,*

LERON – Low Energy Radiation Originar Node,

Λ – *cosmogonical constat,*

NULLON – central inergetic field – energy dual converter (Perfect Black Body/Perfect White Body),

$\underline{R}_i$ - *the vector of each geodesic of Universal expansion,*

STEMION - Space, Time, Energy, Mass, Information Originar Node,

T_{ij} - *energy stress tensor,*

The Unique Force Pattern

$\mathcal{T}$ – *prenergetic period (total time of a inergy - energy/energy – inergy transformation)*,
Θ - *tetus - the operator of the infinite-dimensional assembly:*
$ü, ů, ū$ - *the relative position between two particles,*
$z \, \Theta \, y = \{ -izy..., -zy, ... \, 0, ... \, zy, ... \, izi\}$.

The theoretical approach through which is defined and explained this pattern can still be found in the Romanian complete study, by the same author.

STEMIONICS

ANNEX 1

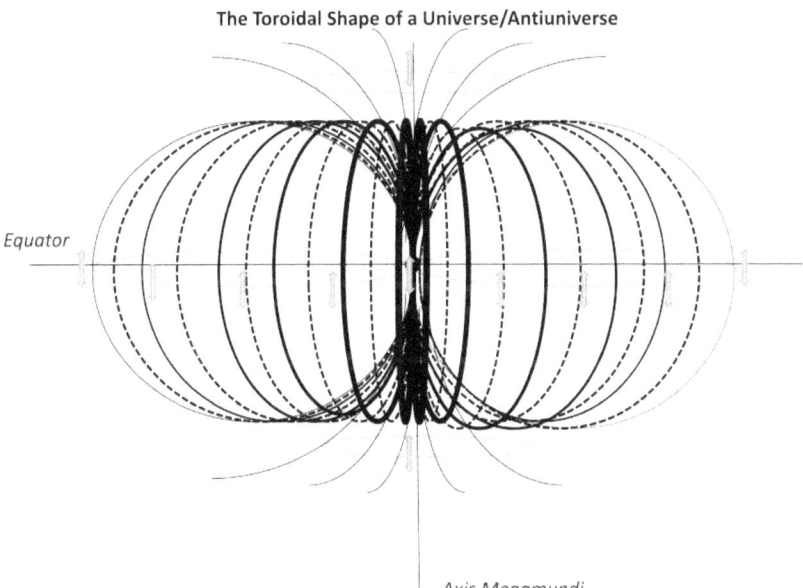

STEMIONICS

The Unique Force Pattern

ANNEX 2

The Transport Energy geodesics

Equator   NULLON   Potential Dump

Axis Megamundi

STEMIONICS

ANNEX 3

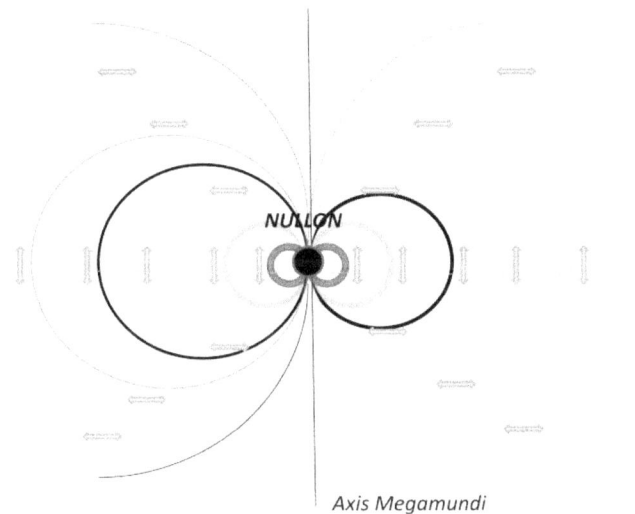

… STEMIONICS

ANNEX 4

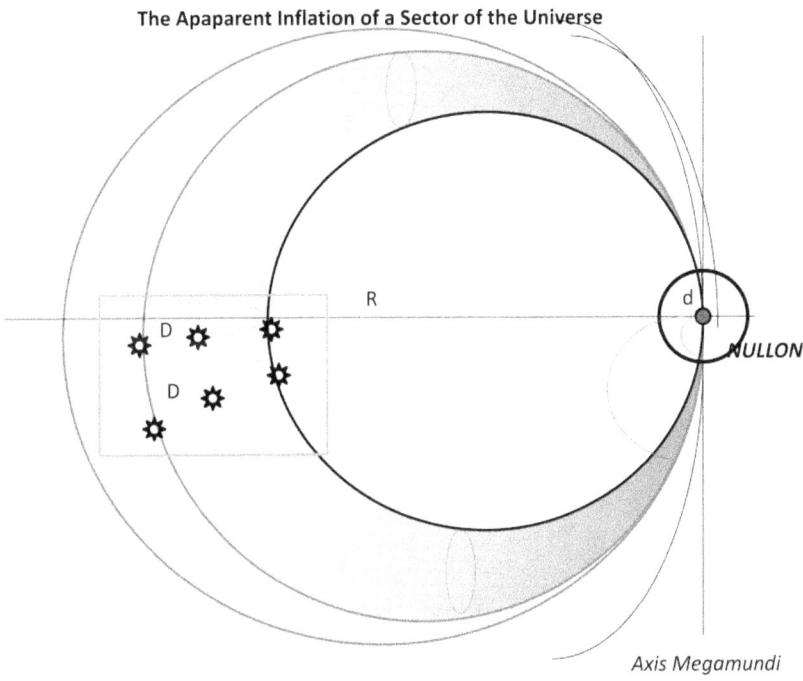

STEMIONICS

ANNEX 5

The Continuous Potential of The Energetic Quanta

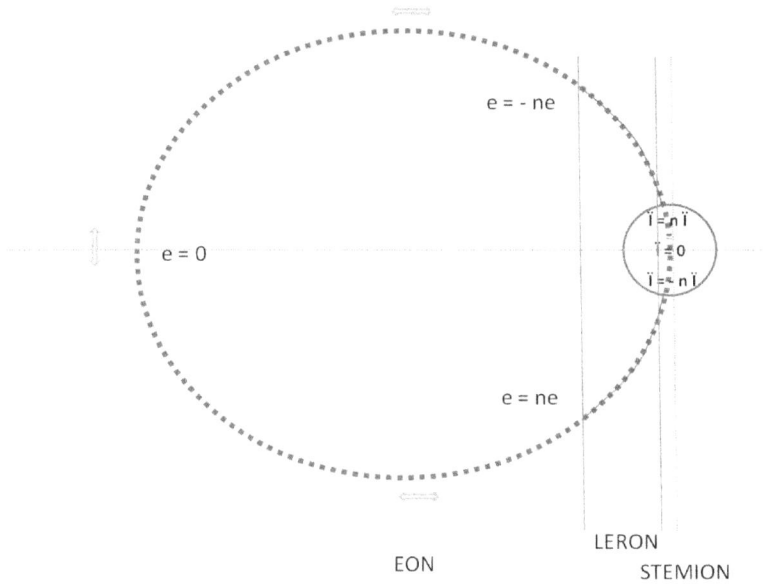

EON
LERON
STEMION

STEMIONICS

The Unique Force Pattern

INDEX

The Story of Stemionics	5
The Objectives of the Pattern	7
The Standard Stemionic Pattern of Grand Unified Theory	9
Annex 1	33
Annex 2	35
Annex 3	37
Annex 4	39
Annex 5	41
Index	43

STEMIONICS

www.ingramcontent.com/pod-product-compliance
Lightning Source LLC
Chambersburg PA
CBHW070339190526
45169CB00005B/1959